SPRINKLER SYSTEMS EXPLAINED

A guide to sprinkler installation standards and rules

bre press

breglobal

The BRE Group is the UK's leading centre of expertise on the built environment, construction, energy use in buildings, fire prevention and control, and risk management. BRE Global is a part of the BRE Group, a world leading research, consultancy, training, testing and certification organisation, delivering sustainability and innovation across the built environment and beyond. The BRE Group is wholly owned by the BRE Trust, a registered charity aiming to advance knowledge, innovation and communication in all matters concerning the built environment for the benefit of all. All BRE Group profits are passed to the BRE Trust to promote its charitable objectives.

The BRE Group is committed to providing impartial and authoritative information on all aspects of the built environment. We make every effort to ensure the accuracy and quality of information and guidance when it is published. However, we can take no responsibility for the subsequent use of this information, nor for any errors or omissions it may contain.

BRE, Garston, Watford WD25 9XX
Tel: 01923 664000
Email: enquiries@bre.co.uk
www.bre.co.uk

BRE publications are available from
www.brebookshop.com
or
IHS BRE Press
Willoughby Road
Bracknell RG12 8FB
Tel: 01344 328038
Fax: 01344 328005
Email: brepress@ihs.com

Requests to copy any part of this publication should be made to the publisher:
IHS BRE Press
Garston, Watford WD25 9XX
Tel: 01923 664761
Email: brepress@ihs.com

Acknowledgements

Our thanks to all members of the LPC Fire & Security Board and affiliated companies and organisations for comments and advice, and to Frontline Fire International Ltd for initial drafting work.

Grateful thanks are extended to the following for permission to use photographs in this guide.
Day Impex Ltd: Figure 3
Hall & Kay Fire Engineering Ltd: Figure 11
Frontline Fire International Ltd: Figures 12–14, 16, 18–22, 24
Reliable Sprinklers: Figure 23

CONTENTS

ACRONYMS

AMAO	assumed maximum area of operation
BMS	building management system
EPEC	enhanced protection extended coverage
ESFR	early suppression fast response
FOC	Fire Offices Committee
FPA	Fire Protection Association
HTM	health technical memoranda
LPC	Loss Prevention Council
LPS	Loss Prevention Standard
RTI	response time index

FOREWORD

We often face the unexpected, whether it is a delayed flight, a cancelled appointment, or a letter from the 'local safety camera partnership'! While most unexpected events do not have a dramatic effect upon our lives, fire does. In almost every instance significant damage to property is the result, or worse still people are killed or injured. Fires continue to make the headlines almost every day, yet the instance of fire is still unexpected.

Based on over 100 years of events and evolution the standards upon which sprinkler systems are designed and installed seek to prevent the unexpected event. Almost every eventuality is addressed to make sprinkler systems as reliable as possible. Key components need to be fully approved, the installation designed and installed by competent and approved installers, and the system maintained by approved service engineers. Page by page the *LPC Rules for automatic sprinkler installations incorporating BS EN 12845 (LPC Rules)* offer prescriptive design instruction which, before a sprinkler system can be certificated must be rigorously adhered to by installation designers and installers. Such exact standards are sometimes seen to be excessive and inflexible, but the history of real events upon which the standards are based should never be ignored or the consequences could be unexpected.

This guide is intended as an aid to understanding sprinkler installations and the rules to which they are designed. I hope you find it useful.

Philip Field
Technical director
LPCB

1   INTRODUCTION

The purpose of this guide is to give some background and commentary on the purpose and uses of sprinkler systems and to explain their important features clearly. It will explain some of the engineering behind the standards and regulations which apply and clear up some of the common misunderstandings about sprinkler systems. It is not intended to be a design guide; the correct implementation of the standards called for in the *LPC Rules for automatic sprinkler installations incorporating BS EN 12845*[1]† (the *LPC Rules*) should be entrusted to individuals with appropriate levels of qualification, training and experience.

There is much mystique associated with sprinkler systems; sometimes the *LPC Rules* are seen by those outside the sprinkler industry as difficult to understand and too rigid in application. It is true that the parameters of sprinkler design are often applied as set in the stone – and it can be frustrating for other building services disciplines when those rules cannot be manipulated to suit a particularly difficult problem or similar issue. The reality is that with almost all other building services, their function can be fully proven during the commissioning stage and any shortfall in capability addressed at that time.

Sprinkler systems cannot be fully tested in their working situation because, of course, the ultimate test of the design will only occur in a very extreme set of circumstances of a fire in a building and these cannot normally be replicated at commissioning stage. It is inevitable, therefore, that extreme caution is used when considering if adjustments (known as non-compliances) to the requirements stated in the *LPC Rules* can be accommodated. It is usual for the persons ultimately responsible for accepting such changes (possibly the local authority or fire insurers) to be consulted prior to the event – time constraints can often lead to a 'stick to the *LPC Rules'* policy. That is not to say that adjustments cannot be made but these should be based on sound engineering principles, erring on the side of caution where there is doubt, and be discussed with all relevant parties prior to implementation.

Although other international standards are discussed in this guide, the principal guidance here relates to the *LPC Rules*.[1]

BS EN 12845:2004[2] is able to stand alone but the *LPC Rules* include BS EN 12845 (2003 edition), as well as a series of technical bulletins which supplement the core document. Where there is conflict, the LPC technical bulletins would take precedence. The technical bulletins are owned, published and maintained by the Fire Protection Association (FPA). Before 2003, when BS EN 12845 was introduced, the core document used in the *LPC Rules* was the 1990 edition of BS 5306-2 *Fire extinguishing installations and equipment on buildings – Specification for sprinkler systems*.[3]

Sprinkler protection should not be considered in isolation but should be part of a package of measures used throughout premises on how to deal with fire-related issues. Indeed, there may be significant interaction between elements of the measures, such as the sprinkler system and the fire alarm and detection system, which are vital to the management of any fire incident. Good fire awareness and management of fire risks and practices will serve to reduce the likelihood of the operation of the sprinkler system.

The possible effects of sprinkler activity on the site, and such factors as firewater run-off, must be carefully considered. The environmental effect of contaminated firewater running to water courses is one of the issues which should be included in the process. That is not to say that the presence of a sprinkler system will increase the likelihood of problems with run-off. Indeed, the fact that sprinklers are likely to act promptly to control an outbreak of fire may reduce the likelihood of very large firewater discharges taking place when the fire and rescue service (FRS) deploys its firefighting jets onto an established fire.

---

† The *LPC Rules for automatic sprinkler installations incorporating BS EN 12845* refers to the 2003 edition of *BS EN 12845: Fixed firefighting systems – Automatic sprinkler systems – Design, installation and maintenance*. This edition has now been superseded by the 2003 edition of BS EN 12845, but the *LPC Rules* have not been updated accordingly because there are no material differences between the two versions.

2 HOW A SPRINKLER SYSTEM WORKS AND WHAT IT IS EXPECTED TO DO

The vast majority of sprinkler systems that are installed are very simple in design and operation. There are some methods of making the control of sprinkler systems more sophisticated, and these are discussed later in the guide, but essentially, they are very straightforward. They comprise of:

- a water supply, which may be the town mains in the street, where this is adequate in capability, or a water tank and pump or similar self-contained water source
- a means of controlling the supply of water into the sprinkler system, usually comprising sets of valves
- a network of pipes which feeds water from the supply to the sprinkler heads
- an array of sprinkler heads, strategically placed throughout the protected area.

It should be appreciated that this is a simplistic representation of the principles (Fig. 1).

It should be noted that in a typical sprinkler system only the sprinklers closest to the fire will be brought into operation. This is because each of them includes an individual heat detection device, usually a glass bulb but sometimes a fusible metallic element, which will operate in relation to the local thermal conditions. It is a common misconception that when activated by a fire, all of the sprinklers will operate. This false idea may be perpetuated by film and television dramas where such an effect can be useful to the plot. In reality, a sprinkler system would be designed to activate a large number of sprinklers

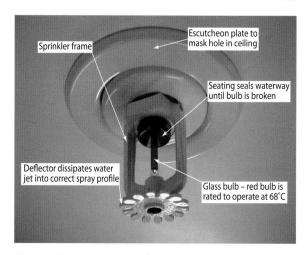

Figure 2: Component parts of a sprinkler head.

almost simultaneously only where there is a risk of a very rapidly growing fire, such as an aircraft hangar. Because the sprinklers are designed to operate individually, a small, relatively slow growing fire may be controlled by a single sprinkler. In the vast majority of fires protected by sprinklers no more than four sprinkler heads are brought into operation, although the design would allow for a larger number than this.

The component parts of a typical ceiling-mounted sprinkler are indicated in Figure 2. The method of actuation of the sprinkler is that the gas temperature close to the sprinkler head will be increased as the adjacent fire grows in size. This increase in gas temperature will

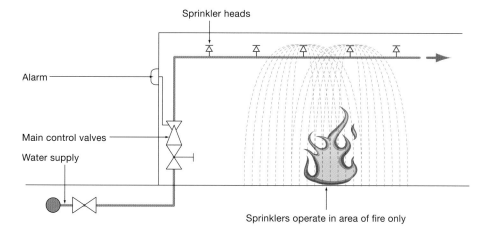

Figure 1: Simplified diagram of a sprinkler system.

progressively increase the temperature of the glass bulb (or metallic link on some types of sprinkler) until its operating temperature is reached. At this point, the bulb is designed to fail and release the seating which allows water to discharge onto the deflector – this converts the water stream into a particular spray pattern.

The temperature at which the sprinkler is designed to operate is usually set to give a margin of 30°C between the maximum expected ambient temperature and the sprinkler rating. This normally results in a sprinkler rating of 68°C and the familiar red coloured bulb (other temperature ratings have different coloured bulbs). Although the rated temperature of the bulb is fixed at a specific temperature, the time at which the first sprinkler will operate will depend upon a number of factors such as:

- the rate at which heat is being released from the fire. This will generally be dictated by the combustibility of the material on fire and how this is arranged or stored
- the height of the ceiling above the fire where the sprinkler is located
- the sprinkler bulb temperature rating and the ambient temperature prior to the fire starting
- the type of sprinkler and its sensitivity rating[‡]
- the distance of the sprinkler head/bulb below the ceiling.

Modern sprinklers are grouped into three categories: quick response, standard response and unrated.

Although the construction of the sprinkler frame will affect the speed at which the sprinkler operates, the most important factor in the determination of sprinkler sensitivity is the size of the bulb. Generally the narrower the bulb the quicker it will react to the local thermal conditions and over the years bulbs have been engineered with smaller diameter bodies and, consequently, faster reaction times. Some samples of various sized and temperature rated bulbs are indicated in Figure 3.

The unrated sprinklers have not been tested for response time and, therefore, have no RTI (response time index). Usually it is only necessary for specifiers to call

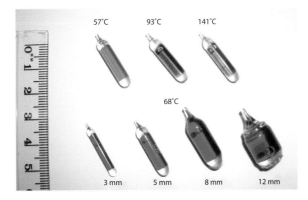

Figure 3: Typical sprinkler bulbs.

‡ During the process for approval of sprinkler heads they are tested to establish their response time index (RTI), which is a reflection of the speed at which the bulb will react to the local hot gases.

§ Flashover is defined as the rapid transition from a growing to a fully developed fire.

for the grouping of sprinkler type rather than the RTI. For instance, if a sprinkler system is being provided for life safety purposes it would be usual for 'quick response' sprinklers to be specified.

The RTI is only significant if accurate predictions of sprinkler response time are necessary when, for instance, a fire engineer is calculating the volume of heat, smoke and other combustion products which need to be considered in a design for smoke removal during a fire. They do, however, give us an opportunity to consider the effect of sprinkler response speed in combination with other factors on the likely time of sprinkler activation and the likely size of the fire when this takes place.

Figure 4 indicates the approximate time for a sprinkler to activate at various heights and with three different fire growth (heat release) rates. This illustrates the combined effects of ceiling height and the speed of fire growth on the expected time for sprinkler activation. It could be considered that the three fire growth rates of slow, moderate and fast could relate to the type of risk present in typical art gallery, office and retail facility respectively.

For example, in an office occupancy with a ceiling height of 3 m using quick response sprinklers, sprinkler activation would be expected in around 190 seconds.

Another important principle to understand is the effect which can be expected when sprinklers are brought into operation (Fig. 5). The minimum effect which can be expected is control of the fire. That is to say that the heat release rate is held to a level that was achieved prior to the operation of sprinklers. In these cases, the final extinction of the fire would be accomplished by the attending fire fighters. However, it is possible that the effect of sprinklers will be greater than the minimum and that the heat release rate will be progressively reduced, even to the point of extinction.

When sprinklers are used as part of a fire-engineered solution within a building, it is conservatively assumed that the fire size will not continue to grow after the first sprinkler has operated and that the heat release rate will be, at most, constant beyond that point in time. It is also to be expected that as the smoke layer will be cooled by the operating sprinklers to below 100°C, flashover[§] is not likely to occur.

One further important aspect is the degree to which the fire has escalated before the first sprinkler operates. A common misconception is that sprinklers will operate at the first sign of fire. The reality is that the fire must develop to release sufficient heat to raise the temperature of the gases close to the sprinkler head to a point at, or beyond, the sprinkler operating temperature prior to the first sprinkler operating. The usual method of describing the fire size is the heat release rate and this is usually expressed in kW (or for larger values MW). This may not mean much to many people, and different materials may give a different visual impression, even if they have similar heat release rates. By way of example, an old upholstered office chair could, in a fully developed fire, contribute a peak heat release rate of 700 kW or 0.7 MW.

It should be borne in mind that the expected outcome of sprinkler activity (indicated in Figure 5) will only commence when the water is discharged from the

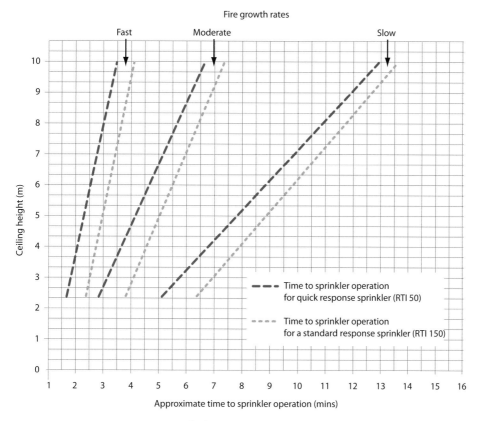

Figure 4: Approximate operating speed of sprinklers.

sprinkler. If a sprinkler system is charged with air, as it would be in a dry or alternate wet and dry system (see section 7), the delivery of water may be delayed. A delay of up to 60 seconds may be expected between sprinkler activation and water delivery in such systems. The heat release rate at the time the control process commences will, consequently, be greater.

Figure 6 indicates the calculated heat release rate at sprinkler operation using the three different heat release curves described in relation to Figure 4. Note that the highest ceiling height shown on Figure 6 is 10 m. However, in a warehouse or factory a sprinkler system is often fitted at much higher ceiling heights. In an atrium, the generally accepted maximum height for ceiling operated sprinklers is 20 m. Using the same calculation process as was used for Figure 4, a sprinkler operation time of over 14 minutes and a heat release rate of 8.67 MW is predicted if the fire is on the floor. However, in practice a fire this size could only be produced by a large object, in which case material would be burning at a high level above the floor.

Figure 7 shows the relationship between heat release rate and the point at which a smoke detector and sprinklers of different response types activate in a typical office with a 3 m ceiling. It should be emphasised that the figures given in these examples are theoretical and, although based on practical tests, some extrapolation is inevitable. In reality, the behaviour of fires may be significantly influenced by simple matters such as the natural ventilation of a space or the quality of risk control within premises.

The *LPC Rules* are drawn up to withstand some of this unpredictability, with such issues as the anticipated number of operating sprinklers built into the design being well beyond those which emerge from statistics. Another

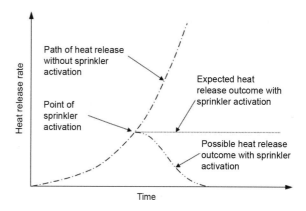

Figure 5: Expected effect of sprinklers on heat release.

issue is the size of water storage tanks being sufficient to sustain a sprinkler system beyond the time when we would reasonably expect the FRS to have dealt with any final extinction required.

It is sometimes considered that the *LPC Rules* overspecify the requirements of sprinkler systems, which makes them more expensive than they need to be. However, there must be some resilience built into standards to counter, as much as possible, the effects of the unknown or unexpected in the protected building. If, for instance, goods are stored too high or some obstruction is placed between the sprinkler head and the risk, there is the potential for a fire to be much larger and more challenging than it should be for the theoretical hazard classification allocated.

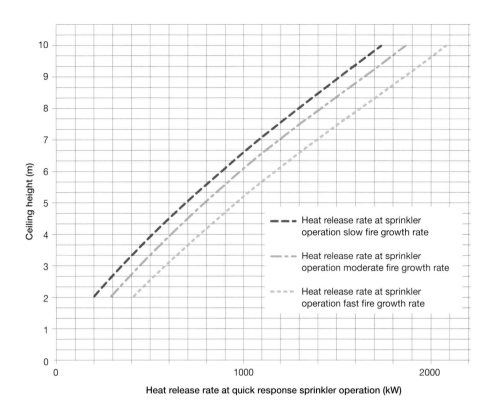

Figure 6: Relative heat release rate at sprinkler operation.

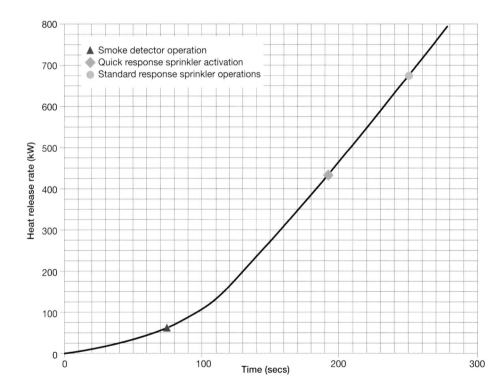

Figure 7: Relationship between heat release rate, time and activation of devices (in a 3 m high ceiling in office occupancy).

3   WHY SPRINKLERS ARE INSTALLED

Sprinklers can be installed for a number of reasons; typically these can be grouped under the following headings.

3.1   TO MEET BUILDING CONTROL OR OTHER AUTHORITY REQUIREMENTS

Some buildings need to be fitted with automatic sprinklers as part of a package of provisions to comply with the Building Regulations. They may be applying local acts or national guidance such as Approved Document B Volume 2,[4] which gives a prescriptive solution to the requirements of the Building Regulations. Usually the need or otherwise for sprinkler protection will be determined by the occupancy type, floor area within a compartment, compartment volume or building height. Some of the guidance may give a choice between situations with and without sprinkler protection where greater structural fire resistance must be incorporated if sprinklers are not provided. These may be considered as compensatory features and the choice, inevitably, will be made based on the effective use of the space, flexibility in the layout of the building, construction cost and similar issues.

Some premises have their own specific features and challenges – important among these are healthcare premises. For these premises, a separate set of guidelines applies entitled 'health technical memoranda'. They are published by the Department of Health in *Firecode – fire safety in the NHS*.[5] In a similar fashion to the Building Regulations Approved Document B,[4] concessions on aspects of such factors as compartment size and fire resistance values are given where sprinklers are fitted.

Wherever sprinkler protection is required by the local authorities, or as part of a scheme where other fire safety measures are dependent upon the presence of sprinklers, this must be considered as required for life safety purposes.

3.2   TO MEET INSURERS' REQUIREMENTS

A fire insurer may call for sprinkler protection to be installed to reduce the risk of a major loss. This loss may be the physical loss of the premises and its contents, and it may include the costs of the interruption to the business itself. In some cases, a reduction in premium may be offered as encouragement to invest in the protection.

The savings made on insurance premiums over a period may eventually lead to a situation where the sprinkler system has paid for itself and the end user can then enjoy the profit of having made the investment. Some risks are sufficiently substantial and the potential for loss so high that an insurer may not be prepared to underwrite the risk unless a properly engineered sprinkler system is installed. The higher risks tend to be those where goods are stored; although some goods present more of a challenge to sprinkler protection than others, it may generally be considered that the higher the goods are stored the greater the potential for a significant loss.

The insurer will normally have a very clear idea of what scope and standard of sprinkler protection is required to adequately protect the risk. Where sprinkler protection is needed only by the insurers it is likely to be considered, as required, for property protection purposes only.

3.3   TO MEET THE RISK MANAGEMENT REQUIREMENTS OF THE BUSINESS/ PROPERTY

A business may have a corporate policy to provide sprinkler protection for its premises as a matter of course because it has recognised the potential damage to its business that a significant fire could cause. These tend to be the larger companies where the benefits of sprinklers are widely accepted, but the policy can work just as well with a small, single premises company.

Some areas of a property may be so important to the continuity of the business that every possible measure must be put in place to prevent a small fire from escalating into a major loss. Some businesses never recover from the effects of a fire even if they are properly insured because major customers are forced to find alternative suppliers and are lost forever.

A fire risk assessment must be carried out for all premises, other than private homes, under the requirements of the Regulatory Reform (Fire Safety) Order 2005. The risk assessment may identify some hazards for which the logical means of protection is automatic sprinklers. It is worth remembering that an automatic sprinkler system will not only activate to control or extinguish the fire but will also give an alarm – this can be considered as a contribution to the fire detection and warning regime within the premises.

3.4 PROPERTY PROTECTION VERSUS LIFE SAFETY

References are often made to sprinkler systems being for property protection or life safety, but whatever the ultimate purpose of a sprinkler system, the bulk of the design will be identical. A property protection sprinkler system gives some life safety benefits and protects the property. The spacing of sprinklers, sizing of pipes and general arrangement of these systems are likely to be identical. The enhancements which must be included in a life safety system mainly relate to further increases in the already high reliability of sprinkler systems; this is to make absolutely certain that water is always available at the sprinkler heads. These include:

- the sprinkler system must be a wet type
- sprinkler systems must be arranged in valved zones with limited numbers of sprinklers in each zone
- all stop valves between water source and sprinkler head must be electrically monitored
- water flow must be electrically monitored
- certain types of sprinkler head are not permitted. It would be normal to specify quick response sprinklers with the appropriate RTI index for a life safety system
- there must be superior (according to BS EN 12845:2003[2]) water supplies.

Although the additional features are attributed to life safety systems, they can be applied equally to property protection systems where the importance of the system is high.

3.5 DOMESTIC AND RESIDENTIAL SPRINKLERS

An area of increased focus is the protection of domestic and residential risks. As the greatest number of lives lost in fires in the UK are in residential premises it follows that the most potential to save lives is in the same area.

A properly designed and installed sprinkler system can operate automatically during the early stages of a fire.

The system could not only could control and potentially extinguish the fire, but vitally limit the release of combustion by-products, such as smoke and toxic gases, which can be so harmful to the residents of the premises. This is even more important when the occupants are very young, elderly or disabled, for whom rapid escape may be difficult or impossible.

The special provisions for such protection are contained as supplementary information in the *LPC Rules*.[1] In addition to the *LPC Rules*, guidance on protection is given in the code of practice BS 9251:2005.[6]

An LPCB scheme exists for the approval of installers of residential and domestic sprinkler systems, namely LPS 1301 *Requirements for the approval of sprinkler installers in the UK and Ireland for residential and domestic sprinkler systems*,[7] which is discussed in section 13.

3.6 SPRINKLER PROTECTION OF SCHOOLS

The recognition of the ability of sprinkler systems to provide a reliable and resilient method of property protection for schools has led to the development of specialist guidance alongside the main guidance provided within the *LPC Rules*. Technical bulletins in the *LPC Rules* have been added and these focus on the classification of hazard, selection of sprinklers and provision of water supplies, and the requirements which have been adjusted in line with specialist area needs. The guidance recognises concerns in respect of vandalism to sprinklers and although it is never possible to have an effective automatic firefighting system which is not vulnerable to some degree, the use of concealed type sprinklers (discussed in section 9) is permitted in certain circumstances so that the presence of sprinklers can be less obvious.

The Department for Children, Schools and Families has published a guidance document for the provision of sprinklers in schools titled SSLD 8 *Sprinklers in schools, Standard specifications, layouts and dimensions*.[8]

4   EXTENT OF SPRINKLER PROTECTION

One of the main principles of automatic sprinkler protection is that it is geared to detect and control a fire in its initial stages, rather than to deal with an established blaze. It is essential, therefore that, in general, the entire premises are sprinkler protected.

Some areas are considered as necessary exceptions to sprinkler protection. These are areas where the operation of sprinklers may present a hazard in itself or cause some adverse reaction between the water and other materials. Some areas are considered as permitted exceptions and these will generally be those where the fire load is very low or where another automatic extinguishing system is installed. In any case, when sprinklers are not installed it is important that the fire resistance of the construction between sprinklered and non-sprinklered areas is sufficient. The minimum value of the separation is specified within the *LPC Rules* and it is important that any openings in the separating constructions are adequately considered. For

instance, doors should be self-closing or automatically closed in the event of fire and any penetrations for building services should be properly sealed.

An example of how this may work is indicated in Figure 8. The sprinkler-protected areas are highlighted in yellow. Non-protected areas are shown in blue and in each case, the passive separation between protected and non-protected areas must meet the requirements of the *LPC Rules*. The green area represents a space where another form of automatic fire protection – in this case a gaseous total flooding system – is provided and sprinklers are not included.

The areas outside of the buildings on the premises should not be overlooked when considering the sprinkler protection. Such factors as idle pallet storage and nearby risks which could form an exposure hazard may need to be catered for in the overall fire management scheme, of which the sprinkler protection will form part.

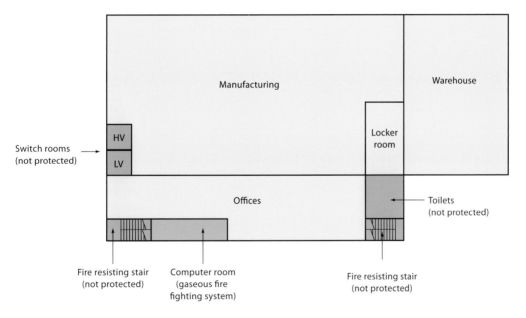

Figure 8: Example of extent of sprinkler protection.

5 HAZARD CLASSIFICATION

Appropriate classification of the risk and hazard is vital if sprinkler protection is to provide the quality of protection expected. It may be anticipated that sprinkler systems designed and installed to the *LPC Rules* will cope well with minor changes to the risk such as limited areas where the goods are stored slightly higher than planned for. However, it should be presumed that if the hazard classification is not properly assessed, or the occupancy is changed to make it beyond the original design, it is possible that the sprinkler system will not control an outbreak of fire and a loss could occur.

Occupancies are identified in separate groups and some of these are further sub-divided to give refinement for the specification of system requirements. These general groups are classified as light, ordinary and high hazard. Each separate classification and sub-classification will define three important target values:

- the design density, expressed in mm/min. (L per m² per minute)
- AMAO in m²
- the duration of the supply (in the case of stored water supplies).

Essentially, the design of the sprinkler system will be geared to accommodate the simultaneous operation of the number of sprinkler heads located within the AMAO (assumed maximum area of operation). The sprinkler system will be hydraulically designed to ensure that at least the minimum flow of water from each sprinkler head will provide the design density in that area. The hydraulic load of this design is built into the design for the water supplies, which will be geared to be sustained for the operating duration.

By far the most commonly used of these is ordinary hazard group 3, and in a wet system designed to this standard the design density is 5 mm/min. and the AMAO is 216 m². What this means is that in the area where the sprinklers operate (this equates to at least 18 sprinklers in an ordinary hazard group 3 risk) the rate at which water would be delivered from the sprinklers would average at least 5 mm deep for each minute of operation. If the sprinklers operate for 20 minutes, then at least 100 mm of water would have been laid down below the operating sprinklers. The water not consumed by the fire (by evaporation) or retained in the goods would, of course, run by gravity to the adjacent lower points in the building. The water may also run outside the building, as well as wetting the immediate area below the sprinklers.

It should be borne in mind that there will be specific limitations on a sprinkler system with that specification, particularly with regard to the maximum height to which goods can be stored, and an ordinary hazard group 3 system, although the most common, should not be regarded as the universal sprinkler system. It is vital that the risk is properly assessed and an appropriate classification apportioned.

There is very good guidance in the *LPC Rules* on hazard classification. Flow charts and lists of typical occupancies are included in the guidance. However, it is essential that experience and judgement is used when undertaking this vital task because there are additional considerations such as the size of rooms, processes, methods of goods storage and packaging of products which must be taken into account. The content of plastics within the products, as well as packaging and use of combustible materials for such things as pallets and tote boxes, figure significantly in the method of assessment for storage risks.

An example of a specific set of storage limitations is shown in Figure 9. These apply to a situation where the protection has been designed to ordinary hazard group 3 and where the goods have been classified as category 2. Higher categories of goods would impose more limited storage.

There is maybe more than one classification in one premises. For instance, a manufacturing facility may include a high hazard process risk in the manufacturing areas, an ordinary hazard risk in the offices and staff areas, and a high hazard, high piled storage risk in the goods-in and warehouse areas. The designs for each area would be geared to suit the individual risks but it is likely that the water supplies would be centralised and geared to the most demanding risk.

Figure 10 indicates how this may work in practice. The wet sprinkler protection is shown in yellow and the alternate wet and dry protection in orange.

The jurisdictional authorities, that may include the local authority building control departments and fire insurers, are likely to have indicated their preferences in terms of hazard classification. If there is confusion or doubt, it is best to seek assistance from an LPS 1048-approved contractor or competent specialist consultant for professional guidance.

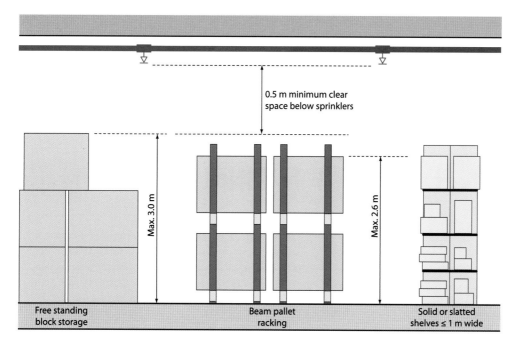

Figure 9: Ordinary hazard group 3 storage limitations.

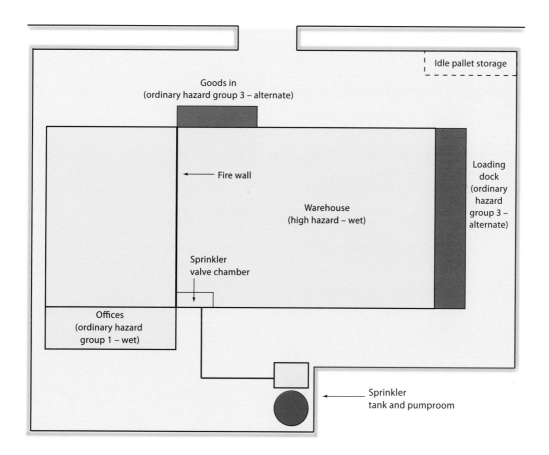

Figure 10: Hazard classifications on one site.

6   SPECIAL CLASSES OF SPRINKLER SYSTEM

6.1   EARLY SUPPRESSION FAST RESPONSE

A special type of protection for high hazard storage risks has been developed. Unlike any other normal sprinkler system, this high performance style is designed to suppress or reduce the size of the fire rather than just control it. It is termed early suppression fast response (ESFR – see Fig. 11) and, as the name suggests, relies upon the earliest possible deployment of water onto the fire in quantities sufficient to extinguish the outbreak rapidly. Because of the high flow rate of water required, the sprinkler heads have very large inlet bores and the pipe sizes tend to be larger than normal to accommodate the flows involved. The main aim of an ESFR system is to avoid the use of in-rack sprinklers, which may be difficult to instal or impose limitations on the flexibility of the logistical use of the facility.

The design of ESFR systems is covered in a technical bulletin of the *LPC Rules*.

Only certain risk types and building configuration are suited for ESFR sprinkler protection. For instance, rolls of tissue paper cannot be protected with ESFR sprinklers because of the rapidly growing fire rate. In this situation, traditional sprinkler protection with in-rack sprinklers would need to be deployed. The storage methods and configuration are very important to the effectiveness of the protection and very specific parameters are laid down in the *LPC Rules*.

Another important prerequisite to the use of ESFR sprinklers is the slope of the roof. If the roof slope is steep the hot gases, which will activate the sprinklers, will tend to run rapidly up the slope and for this reason a maximum slope of 9.5° is imposed. There are many other specific requirements written into the technical bulletins and these are mainly concerned with achieving the early and accurate activation of the sprinklers in a fire.

Figure 11: ESFR protection of large high hazard risk (prior to occupation).

6.2   ENHANCED PROTECTION EXTENDED COVERAGE SPRINKLER PROTECTION

The latest innovation in sprinkler design technology is the enhanced protection extended coverage (EPEC) system. This is for use in ordinary hazard group 3 risks. The principal concept is that special sprinkler heads are used which have been developed specifically for this type of system. The heads are spaced at greater distances and areas than a normal sprinkler system. The EPEC system is designed in accordance with a separate technical bulletin and, similar to the ESFR approach, the principles given in the *LPC Rules* should be rigidly applied.

Only certain risk types and building configuration are suited for EPEC sprinkler protection. The maximum permitted roof slope is 9° and the maximum ceiling height is 5.5 m. There are limitations on the type of goods stored and, like the ESFR systems, goods which give rise to unusually severe fire characteristics need to be avoided.

7 TYPES OF SPRINKLER SYSTEM AND CONTROLS

There are many options available to designers of the arrangement used to control the sprinkler systems. These would normally take into account the circumstances of the risk, particularly in connection with the expected temperature within the building, but in some cases to deal with other issues. The best approach is to keep the sprinkler system as simple as possible and to avoid any complications whenever it is practical to do so.

The common theme with all sprinkler system types is that they will have an arrangement of control valves which will include an isolating valve and an alarm device to indicate when water is flowing. Other components may be included, such as air compressors and control panels, depending upon the type of sprinkler system. The location of these control valves is important because the FRS will need to access them when it attends a fire.

In some high piled storage risks it will be necessary to fit sprinkler protection within the storage racking as well as at the roof level. This is because the fire is likely to be too well established to fight from roof level only in a normal sprinkler system. In most cases, separate installations are provided for the roof and rack systems.

Each type of sprinkler system will have parameters set within the *LPC Rules* for such matters as maximum number of sprinklers fed (or the maximum area to be fed), types of sprinkler permitted, maximum height of building section to be protected and pipework arrangements.

Different types of sprinkler system are discussed below.

7.1   WET INSTALLATIONS

These, the most common of all installations, are designed for use where the installation is in an area which is not subject to frost at any time (Fig. 12). The temperature for this demarcation is usually taken as 4°C. In addition, a wet installation cannot be used where the ambient temperature exceeds 95°C.

The sprinkler system is charged with water at all times and the operation of a sprinkler head will result in the immediate release of water to fight the fire. This is the main reason that a wet installation is the first choice and the immediate issue of water will start the fire control process at the earliest time.

Figure 12: Typical wet installation control valves (with life safety by-pass).

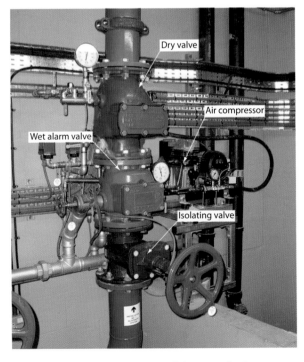

Figure 13: Typical alternate wet and dry control valves.

7.2 ALTERNATE WET AND DRY INSTALLATIONS

Where a risk area is subject to natural frost conditions water-filled pipework cannot be used. To overcome this, an alternate wet and dry installation can be fitted (Fig. 13). The control valves on these systems are designed to allow the system to be filled with water during the months of the year when frost is not likely, and to revert to dry operation when frosts are possible. During the colder months, the system is charged with air at a modest pressure and when a fire activates a sprinkler head, the air is released and water is allowed to enter the system. With larger systems, a device can be fitted at the control valves to accelerate the process of system charging and air release to improve the speed of water delivery to the operated sprinkler. Similar to a wet installation, these systems cannot be used where the ambient temperature exceeds 95°C.

7.3 DRY INSTALLATIONS

Where a wet or alternate installation cannot be used, eg in a cold store where the ambient temperature is permanently maintained below 4°C, or in drying ovens where the temperature is raised above 70°C, a dry installation can be used. The operating principle of dry installations is the same as winter operation of an alternate system described previously.

7.4 TAIL-END ALTERNATE SYSTEMS

If a system of the alternate or dry type is small, with a total number of sprinklers lower than 100, this tail-end can be arranged as an extension to wet or alternate installation.

A typical example of such an arrangement is a small, unheated loading bay at one end of an otherwise heated building. The wet installation feeding the main building could be extended into the loading bay area through a tail-end alternate system.

7.5 PRE-ACTION INSTALLATIONS

Pre-action sprinkler systems are installed in combination with electronic fire detection systems. There are two types of pre-action installations: Type A and Type B.

Type A

Type A systems are intended to prevent the premature release of water from a system which has suffered mechanical damage to the sprinkler head or pipework. Type A systems are often installed in an area where the unwanted release of water would be particularly damaging, such as a data processing hall.

Type A systems are arranged so that the control panel will not allow the opening of the pre-action control valve until fire has been confirmed by the fire detection system. The detection should signal fire in advance of the activation of the sprinkler head, so that the system pipework is charged with water by the time it is required by the sprinkler operation. In its quiescent state, the system pipework is charged with monitored low-pressure compressed air so that any damage to the sprinkler heads or pipework will raise an alarm but not result in water release. A confirmation of fire by the detection system, and the operation of a sprinkler head, is required before water is discharged.

Although type A systems give comfort to those who fear damage to precious equipment from unwanted sprinkler operation, the complication of the interface with a detection system will serve to reduce reliability. This type of system should be used only when there is a real danger of it being damaged, and where no option exists to mechanically protect the vulnerable portions of a wet installation.

Type B

Type B systems are intended to facilitate early discharge from a dry or alternate sprinkler system by releasing the main control valve to charge the system with water upon operation of the fire detection system fitted in the same area as the sprinklers. This is intended to avoid the delay in delivery of water onto the fire which could occur while the system reacts to the loss of air pressure after sprinkler activation. This is likely to be specified when large systems or risks with potentially rapidly growing intensive fires may occur.

The dry or alternate system will still be charged with compressed air in the same way as standard systems. In the event that a sprinkler head is activated, the loss of air from the system would operate the main valve to charge the system with water. This would happen even if no fire detection signal has been received.

With both types of pre-action system, a properly designed fire detection system complying with BS 5839-1[10] or EN 54-14[11] should be installed. For data processing applications the provisions of BS 6266[12] should also be considered.

7.6 DELUGE INSTALLATIONS

Some risks have fire loads that can be expected to produce very rapidly growing and intensive fires that may not be controlled by the progressive opening of sprinklers by the local thermal conditions. Two examples of this would be firework manufacturing and the protection of an aircraft hangar. In the latter case, it is likely that foam would be added to the water in the system to deal with the liquid fuel (Class B) risk. Another situation where a deluge installation would be used is the protection of an external plant area from the radiated energy effects of an adjacent fire. In this case, special directional sprayers or nozzles are used to envelop the risk.

Figure 14 shows an external deluge system under test which provides exposure risk protection to a road tanker unloading terminal. The defined cone of the spray nozzles can be clearly seen.

Figure 14: External deluge protection.

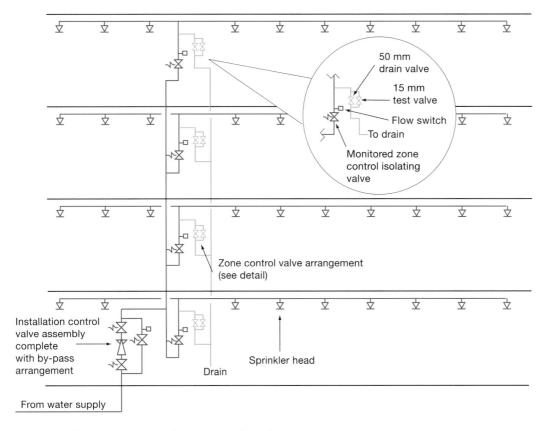

Figure 15: Schematic of typical multi-storey zoned installation.

Deluge systems are fitted with unsealed or open sprinklers and pipework which is empty. As with other sprinkler systems it is controlled by a set of valves and these are designed to hold back the water until called for either by a dry pilot system comprising sprinkler heads and small bore pipework charged with compressed air run in the same area as the open sprinklers, an electronic detection system (see section 11) or manually. The water is then allowed to flow into the sprinkler system and to discharge simultaneously from all sprinkler heads or nozzles.

Because of the type of sprinkler system, and to limit the water supply requirements, deluge systems tend to be divided to cover smaller areas than other types of system. Often a building will be divided into multiple deluge installations to give the best strategic balance of protection. Obviously, a thorough assessment of the potential for multiple installations to activate simultaneously needs to be carried out. Allowances should also be made for the water supplies to cope with this within the scope of the *LPC Rules*.

7.7   SYSTEM ZONES

It is desirable to sub-divide some wet installations into smaller sections known as zones. By doing this, it is possible for one zone to be shut down for refurbishment, or following a fire, while maintaining the protection in other areas covered by the installation. This is most likely to be seen in life safety installations where the *LPC Rules* require the sub-division of installations into zones of not more than 200 sprinklers, but it is also possible with property protection systems. Typical examples of zoned systems would be a multi-storey office block with zones on each floor and a shopping centre with zones feeding each shop unit. A typical simple schematic of a multi-storey zoned system is shown in Figure 15.

Each zone is usually controlled by an electrically monitored stop valve and includes a water flow alarm switch to signal to the alarm system when water is flowing into the zone. The stop valve is monitored to indicate to the alarm system when the valve is not fully open. A drain point must be provided downstream of the stop valve to empty the zone of water when needed and a test connection is provided downstream of the flow switch to test and exercise the flow switch. A typical zone control arrangement is indicated in Figure 16.

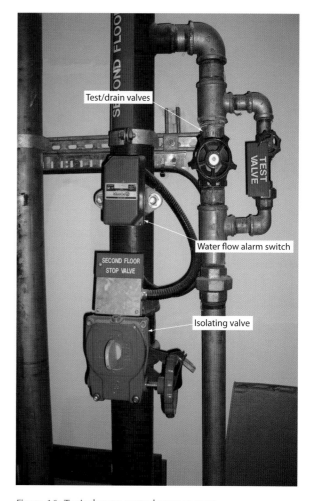

Figure 16: Typical zone control arrangement.

8   WATER SUPPLIES

The supply of water for a sprinkler system is the most important element – without the water at the correct rate and pressure, the system cannot be expected to perform as required. Obviously, some arrangements will be more reliable and resilient than others so a method of grading of the supplies is built into the *LPC Rules*. These take into consideration issues such as:

- The risk in the premises, and the probable time it will take for control to be established and for the FRS to decide that the intervention of sprinklers is no longer needed. A high piled storage risk is likely to take longer to control than an office fire. This will determine the duration of the supply and will convert to tank size if stored water is part of the arrangement.
- The required water flow rate. This will be determined by considering the design density, AMAO and the pipework arrangements (see section 5).
- The required pressure at which water must be delivered into the system. This will be determined by hydraulic analysis of the system demands (see section 10).
- The reliance which can be placed on supplies of water from the local water undertaking. This will determine if these supplies, on their own, can provide sufficient flow, pressure and reliability to satisfy the risk or if they are required to be supplemented or replaced by stored water supplies and pumping arrangements.
- The reliability of mains electricity supplies when electric driven pumps are part of the supply strategy.

These issues are considered in relation to the importance of the sprinkler protection. A life safety system would be expected to have the very highest degree of reliability so it is likely that a supply strategy with a single point of failure would be unacceptable. A property protection risk, where the potential commercial loss is very large, would also be examined in a similar way. On some property protection risks, a higher degree of risk may be tolerable and simpler, more economical water supplies with a lower, but still high, reliability may be acceptable. The decision in such a situation would be a commercial one.

8.1   TYPES OF WATER SUPPLY

Water supplies are graded in three main supply groups: single, superior single and duplicate. These groups describe the type of arrangements and a greater reliance can be placed on a system which is higher up the scale toward a duplicate water supply where higher levels of redundancy are built into the arrangements. It is not intended to elaborate on the combinations which constitute the grading because these are well defined within the *LPC Rules*. In general, the sources which can be considered to contribute to the arrangements include the:

- town mains
- town mains booster pump (where permitted by the water undertaking)
- suction tank(s) and pump(s)
- gravity tank
- elevated private reservoir
- air pressure tank
- pumps drawing from a river or canal (special conditions apply).

Wherever possible it is wise to include an FRS inlet into the sprinkler system so that the supplies can be supplemented when needed. This should take the form of a breaching inlet, similar to that which would be used on a dry riser system, connected by pipework to the sprinkler system. The point of connection to the sprinkler system should be the trunk main which feeds the installation control valves.

It is possible to consider using salt or brackish water as a source, where no other alternative exists, but this water must not be retained within the sprinkler system, which must be charged with fresh water. Arrangements must be made for the thorough flushing of such a system after activation. Similarly, water can be supplied from a swimming pool but it cannot be retained in the system. Other issues, such as pool occupant safety and reliability of the supply, must be carefully considered.

The most common combinations found in UK systems are either direct connections from town mains or a combination of suction tank and pumps (Figs 17 and 18). In the case of town mains supplies, great care should be taken to ensure that these offer a long-term capability of flow and pressure suited to the system needs. The ongoing policy of water conservation has led to reductions in town mains pressures in some areas.

Where two pumps (one duty and one standby) are installed as part of superior or duplicate supply, care should be taken to ensure that power is available when needed, especially during a fire incident. If two electric driven pumps are provided, these need to be supplied from two totally independent electrical supplies. The

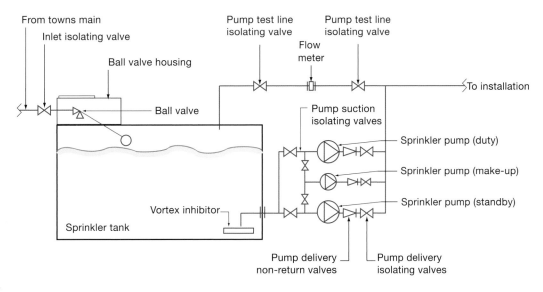

Figure 17: Simplified schematic layout of typical pumped water supply.

Figure 18: Typical arrangement of external suction tank and pump.

Figure 19: Typical multi-stage multi-outlet pump (three stages).

common alternative is to provide one electric driven and one diesel driven unit. Where electrical power supplies are difficult or expensive to provide, it is possible for two diesel driven pumps to be used.

With all water supplies, it is necessary to be able to test the capability, and full flow testing of the supplies must be provided as part of the system. Permission to test town mains supplies must be sought from the water supplier. In the case of stored water, it is often possible to return test water to the tank and avoid wastage.

Where water is stored for sprinkler use there are a large number of manufacturers of different styles of tank from which to select. An approved model of tank should always be used (see section 16). Two general styles of tank are common, vertical cylindrical and rectangular. Common features on all tanks will be a roof, to exclude daylight, and an infill from the town mains, either to replace the contents after use (in a full holding capacity tank) or to contribute to the effective contents while water is being used on the fire (in a reduced capacity tank).

The size, style and number of pumps must be geared to suit the risk and the level of reliability required. The higher grade of risk will almost certainly require higher volumes of water, which will result in the need for more powerful pumps and drivers. Where very large capacities are required, it is possible for the demand to be served by multiple pumps. For instance it is possible to have three pumps fitted, two of which can fulfil 50% of the required demand each, with the third being a standby unit. For very high multi-storey buildings, it may be necessary to feed the sprinkler protection in separate pressure stages in order to limit the maximum height between the highest and lowest sprinkler in each stage, to the maximum 45 m specified in the *LPC Rules*. For instance in a building 135 m high it would be possible to supply the water through a multi-stage multi-outlet pump set with three separate pressure stages (Fig. 19).

9   SPRINKLER TYPES

The development of sprinkler systems can be traced back to the early 19th century. Since then there have been a large number of improvements in terms of aesthetics, performance, reliability and thermal response but the principles have remained largely the same – to deploy water, at the appropriate time in the form of spray, from thermally responsive devices placed on a system of pipework which is permanently connected to a water supply. The devices – the sprinkler heads – have evolved into a large number of different types to deal with specific situations. They all have a screwed thread to attach them to the pipes and a thermal device – a glass bulb or a metallic fusible link or similar connection – which holds a seating in place which seals the waterway. The thermal device will be rated to fail at a particular temperature. When this occurs, the seating is released and water is emitted from the entry orifice onto some form of deflector which distributes the water in the form of spray onto the risk.

Figure 20 shows either end of the evolutionary scale. On the left is a late 19th century sprinkler head – a Witter – while on the right is a modern quick response miniature sprinkler. Both of these will work but, clearly, the older, heavier sprinkler will be much slower to react.

The issues related to fire conditions at the time of sprinkler operation are discussed in section 2. Various types of sprinkler have been developed to act in slightly different ways to deliver the water spray in contrasting styles to meet the needs of the building. It is vital that these are used in the way intended and that those designed for specific orientation, or with specific relationship with building features, are correctly positioned.

Figure 20: Ancient and modern sprinkler heads.

9.1 CONVENTIONAL AND SPRAY TYPE SPRINKLERS

Conventional and spray type sprinkler heads are the most common, modern sprinklers designed for use in most situations, either mounted directly onto the pipes or below a suspended ceiling. They are usually of the miniature type, designed to be as neat as possible to give the least aesthetic impact. The only difference between a conventional and spray type, is the type of spray produced and the direction in which it travels. A conventional sprinkler is designed to direct the spray both upwards and downwards from the deflector in roughly equal proportions. This will produce a significant degree of ceiling wetting as well as direct distribution below. The spray sprinkler is designed to direct the majority of its spray downwards. These sprinklers are designed either for pendant or upright orientation although some are designed so that they can be fitted in either way. These are called universal type.

Figure 21 shows an internal view of a retail store in which conventional pendant sprinklers have been fitted below a suspended ceiling. These are difficult to spot on the image but this does illustrate how unobtrusive sprinklers can be. The sprinklers are all a quick response type.

Where sprinklers are fitted to protect below a suspended ceiling, a rosette or escutcheon plate is usually fitted to mask the hole in the ceiling (Fig. 22).

Conventional and spray sprinklers are manufactured in various orifice sizes to suit the types of hazard so that virtually any type of risk can be protected using these sprinklers.

9.2 CEILING, RECESSED AND CONCEALED TYPE SPRINKLERS

This group of sprinkler types is intended for use on suspended ceilings and they are all designed to give a more pleasing appearance than the standard pendant sprinkler. The designs vary in style from the simple recessing of a standard sprinkler through to those which appear from below to comprise only a flat circular plate.

Figure 21: Conventional sprinklers in a retail site.

Figure 22: Example of a sprinkler head fitted in a false ceiling.

Some of them require the thermal activation of two devices, the flat plate and the sprinkler head which it conceals, before water is released.

The common feature with most of these sprinkler types is that they are likely to operate more slowly than the normal sprinkler. They will be slower to operate because they are closer to the ceiling, which will tend to conduct heat away from the adjacent gases. It is generally accepted that on a flat ceiling the hottest gases are located between 75 and 100 mm below the ceiling. Most ceiling recessed and concealed-type sprinklers will be unrated and, therefore, cannot be considered as quick response.

The aesthetic features of these sprinklers make them more attractive to those concerned with the overall look of the finished ceiling, particularly architects and interior designers, but the impact on the speed of sprinkler response and, therefore, their suitability for use in certain situations, must not be overlooked.

One of the most important issues with concealed sprinklers is that when redecoration is carried out on the ceiling, care must be taken to avoid any contamination of any part of the sprinkler, including the cover plate, with paint or similar substances (Fig. 23). While this also applies to other sprinklers, the risk is greater with concealed sprinklers. Also, it must be ensured that the small gap which exists between the cover plate and the ceiling is not filled with any material. This can delay, or even prevent, operation of the sprinkler.

Figure 24: ESFR type sprinkler head compared with miniature type.

9.3 SIDE WALL TYPE SPRINKLERS

These sprinklers are designed to be located at the edges of flat, smooth ceilings and distribute water primarily in one direction, into the room and onto the walls. They are not a genuine replacement for normal sprinklers because the effectiveness of the spray pattern is likely to be lower and more susceptible to interference than a normal sprinkler and the thermal response of the sprinkler will not be as efficient as one fitted in the conventional location.

They are suited to low corridors or other similar areas where headroom is low. They are often used in hotel bedrooms where no suspended ceiling is fitted in the sleeping area, in which case the sprinkler is fitted close to, and fed through, the bulkhead of the entrance and bathroom area. They should not be considered for high fire load areas and are not permitted in high hazard risks.

9.4 EARLY SUPPRESSION FAST RESPONSE SPRINKLERS

ESFR sprinklers have large orifice entry and quick response elements which have been designed for use in ESFR sprinkler systems only, which are described in section 6.

Figure 24 shows a typical ESFR sprinkler head next to a miniature sprinkler by way of comparison of the size of the head.

9.5 ENHANCED PROTECTION EXTENDED COVERAGE SPRINKLERS

EPEC sprinklers have been designed for use in EPEC sprinkler systems only (see section 6).

Figure 23: Concealed type sprinkler head.

10 PIPEWORK ARRANGEMENTS

10.1   CONFIGURATIONS

Three typical pipe configurations are used in sprinkler systems: a tree, loop and grid:

- A tree configuration consists of terminal pipes downstream of the sprinkler installation valves, branching out into separate mains, and terminating in range pipes that are feeding the sprinkler heads.
- A loop, or a multiple loop configuration, consists of a pipe immediately downstream of the sprinkler installation control valves connecting into a single or multiple loop pipe configuration. Range pipes that feed the sprinkler heads are fed from the loop pipes.
- A grid configuration consists of a pipe immediately downstream of the sprinkler installation control valves connecting into a header, commonly termed the front track. A further header on the opposite side of the building or area is termed the backtrack. Multiple pipes connecting these two tracks are called the cross ties, to which the sprinkler heads are connected.

The principle of a grid pipe configuration is that all of the sprinkler installation pipes contribute to the flow of water to the fire area, therefore spreading the water flow friction losses and subsequently reducing pipe diameter sizes. The three arrangements are shown in Figures 25, 26 and 27.

The *LPC Rules* for the design of sprinkler systems do not provide a basis for pipe diameter sizing for loop or grid systems other than by full hydraulic calculations. Loop, multiple loop and grid pipe configurations must not be used for dry and alternate wet and dry pipe sprinkler installations. A single sprinkler installation may have pipe configurations of all types. In this case, all pipe sizing must be determined by hydraulic calculations.

10.2   PIPEWORK

A sprinkler system can consist of several different pipe materials. All pipework should be easily accessible and not buried in concrete floors.

Underground pipes should have sufficient corrosion resistance. Suitable types of materials for underground pipework are:

- cast iron
- ductile iron
- reinforced glass fibre
- high-density polyethylene.

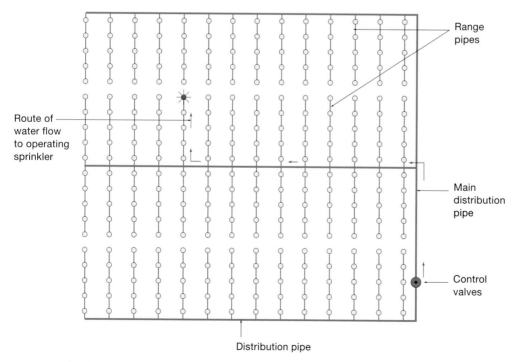

Figure 25: Plan diagram of a tree style pipework arrangement.

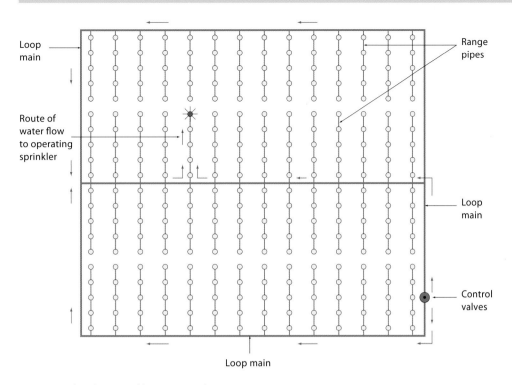

Figure 26: Plan diagram of loop pipework arrangement.

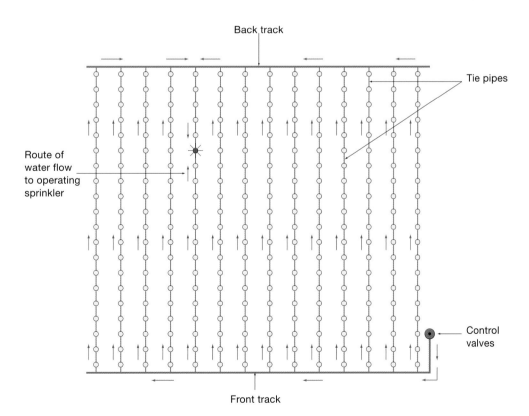

Figure 27: Plan diagram of grid pipework arrangement.

Generally, steel pipework is used for the water supply above-ground arrangement and distribution to the installation control valves. Thereafter, steel, copper, or other materials (in accordance with the appropriate specifications, such as chlorinated PVC) are used downstream of the installation control valves. Please refer to the installation standard for specific pipework approvals and the LPCB Red Book.[13]

The use of galvanised steel pipe is recommended for dry, alternate and pre-action systems.

Sprinkler pipework systems are designed to incorporate the following main features:

- In situations where movement may occur, such as building expansion joints or racking systems, flexible sections or joints are provided.
- Provision must be made to drain all sections of pipework. Auxiliary drains must be provided for sections that cannot be drained through the sprinkler installation main drain valve.
- All pipes must be installed with slopes to drain points.
- Water in pipes that are installed in areas that may be subject to low temperatures must be protected from freezing by the installation of electric trace heating and lagging. (Trace heating will be operated by thermostatic switching.)

10.3   PIPE SIZING

There are generally two methods of determination of pipe sizes in sprinkler systems: pre-calculated and full hydraulic calculation.

With pre-calculated systems, the sizes of pipes at the ends of the system are determined by a simple count-up method whereby a certain number of sprinkler heads can be served by a certain pipe size. There are some stipulations which apply to certain configurations of pipework and the number versus size schedules vary according to the hazard classification and physical arrangements of the pipework. When the number of sprinklers fed reaches a specified limit (the design point) the pipes are sized by a simplified hydraulic calculation method, either to determine the exact pressure required or to stay within stipulated maximum friction losses. This is a somewhat crude but relatively simple method of pipe sizing. The capability of the water supplies feeding this type of system is also predetermined. The main advantage of this method is that subsequent alterations or adjustments to the system may be made without major hydraulic design re-assessments and the capability of the water supplies may be accurately predicted, regardless of the arrangement of the system pipework. It is ideally suited to ordinary hazard risks such as offices and retail situations where alterations and adjustments to the systems are very likely to be required during the life of the building.

The full hydraulic calculation method of pipe sizing involves the detailed hydraulic analysis of the pipework arrays using the specific design density and AMAO values for the risk (see section 5). This is usually carried out by computer and specialist hydraulics software packages are available for this purpose. Multiple calculations may be necessary to establish the most hydraulically unfavourable situation or where the styles of arrays are different. When the water supply is from pumped or similar limited capability sources, the analysis of the most favourable arrays must also be carried out to ensure that the supply has sufficient capability and duration when the hydraulic losses are at their least. The pump capability and tank capacity are dictated by the detailed calculations. This type of pipe sizing method is always used on high hazard risks and on deluge installations. If alterations are carried out to the system, then further calculations must be carried out to ensure that the changed area is correctly sized and that the water supplies are still adequate.

10.4   FLEXIBLE PIPEWORK CONNECTIONS

Two different types of flexible pipework connections are permitted in sprinkler systems:

- Those joining pipework within a high hazard rack storage protection system and the main pipework at the roof.
- Those connecting individual sprinkler heads on a suspended ceiling to pipework located in the void above.

The term 'flexible' is misleading because in both cases there is not an intention for the connections to accommodate significant movement as the term might suggest. In the case of the rack storage, these flexible connections are intended to accommodate minor relative movement between the rack pipework and the roof pipework. In the case of the single sprinkler feeds they are not designed to be flexible in service at all. Flexible pipework connections are used to join the 'first fix' ceiling void pipework to the sprinklers on the ceiling without the need to use multiple short pipe lengths and fittings.

The flexible pipes themselves are subject to third party testing and approval (see section 16) and standards for their installation are identified in the *LPC Rules* and the manufacturers' installation guidance documents. Principal among the requirements laid down are a need to limit the bending radius of the flexible, appropriate fixings and the avoidance of trapped pipe sections.

The *LPC Rules* and the LPCB Red Book[13] contain specific requirements for the installation of flexible pipework connections.

10.5   PIPE SUPPORTS

There are many requirements related to sprinkler system pipe supports, as detailed in the *LPC Rules*. However, the most fundamental of these are:

- A pipe support is to be provided for each section of pipework and must prevent the pipe from sagging, twisting and unnecessary movement.

- Pipe supports have to be fixed directly to the building structure, and only where necessary to machines, storage racks, mezzanines or other similar structures.
- They are to be of the adjustable type completely surrounding the pipe and must not be welded to pipes or fittings.
- All pipe supports are provided for use by sprinkler system pipework only.

10.6   PIPEWORK LIFE EXPECTANCY

Several factors influence the life expectancy of a sprinkler system:

- type of system
- internal finish of the pipes
- quality of the pipes
- frequency of drain-down of the system
- quality of the water
- accuracy of drainage slopes on pipes.

The minimum standards of pipe type are defined in the *LPC Rules* and frequently the choice will be made because of economy, as well as compliance. The use of untreated standard factory finished steel tube is common and although the outside of the pipe may be painted, the internal surfaces are subject to corrosion.

In a standard wet sprinkler system the water may be contained within the system for a considerable time and will be drained fully only when alterations are required. There will be an initial high corrosion rate when the water is highly oxygenated but gradually this rate will reduce as the oxidisation process exhausts the water. In a typical wet system, a life of 50 years is not uncommon. Alternate systems are drained each year and charged with compressed air, then recharged with fresh water for the warmer months. This process encourages higher corrosion rates and for this reason, the life expectancy of alternate systems may be around 20 years.

The use of galvanised pipework will provide corrosion resistance and is likely to extend the life of alternate, dry and pre-action systems (see section 7).

10.7   MICROBIOLOGICAL ISSUES

It is considered that the chances of *Legionellosis* emanating from a sprinkler system are very low but under certain circumstances it is possible, including tanks and pipework which are subject to temperatures over 20°C and the production of aerosols during testing and maintenance. The FPA has published guidance under the technical briefing note *Legionella and firefighting systems*[14] and the recommendations should be observed in all cases.

11 INTERACTION WITH DETECTION AND ALARM SYSTEMS

Sprinkler and deluge systems are most often connected to some form of fire alarm system which monitors the various system accessories.

11.1 DETECTION AND ACTUATION

The detection and actuation of sprinkler and deluge systems are varied. In standard wet, dry and alternate wet and dry sprinkler installations, the sprinkler heads are used for detection. As the elements of sprinklers are heat sensitive devices, sprinkler heads will respond to a rise in ambient temperature. They do not operate on a rate of temperature rise or smoke. Sprinkler heads operate independently of each other and water is only discharged through sprinkler heads that have been activated by a temperature rise to, or in excess of, their specified rating.

Deluge systems are an exception to this rule where open sprinklers or nozzles are installed. Water will discharge simultaneously through all of the open sprinklers or nozzles when the deluge valve is activated.

Electrical/electronic detectors may be used for the detection of fires in areas covered by pre-action and deluge systems. Many systems have the potential to be used as actuation for the sprinklers including:
- heat detector (fixed point)
- heat detector (rate of rise)
- smoke detector (optical)
- smoke detector (ionisation)
- linear detection (fixed point)
- linear detection (rate of rise)
- linear detection (fibre optic)
- beam detection
- flame detection (infrared)
- flame detection (ultra violet)
- aspirating smoke detection
- thermal imaging.

Of these detection systems the most commonly used is for smoke detection. However, sometimes a coincident system is employed where the activation of two detectors is required before the system is released to discharge water.

It is vital that any interaction between a sprinkler system and an electronic detection system is properly designed and engineered, to ensure that the cause and effect is fully considered and that the efficiency of the sprinkler system is not unduly reduced.

11.2 ELECTRICAL SUPERVISION

The electrical supervision of sprinkler systems is recommended, and in cases such as life safety systems and zoned installations, it is mandatory.

Fire alarm systems used to monitor the sprinkler system are activated when water flows through a sprinkler installation. Pressure switches on the sprinkler alarm valve, and flow switches installed in sprinkler risers or downstream of zone control valves, are activated by either water pressure or water flow.

Control valves of either the gate or butterfly design can be monitored in the designated open or closed positions by tamper switches. High and low water levels in firewater storage tank(s) can be detected by float switches or similar. Low air pressure in dry, alternate wet and dry installations, and pre-action installations is monitored by pressure switches. The temperature of pump rooms, water storage tanks and piping with trace heating and lagging may also be monitored with air thermostats.

Fire pump status, faults and pump running indications are provided by direct links to the fire pump controllers/starters. System monitoring for fault or alarm may be through an independent system, or through a fire alarm system, where alarm interface units are provided for interconnection to the sprinkler system devices. It is possible, under certain circumstances, for the building management system to be used to monitor the sprinkler system provided that suitable audible and visual indicators are provided to ensure compliance with the design standard. Whichever method is employed, the integrity of the system and its power supply must be properly arranged.

12 OTHER DESIGN STANDARDS

12.1   RECOGNISED INTERNATIONAL STANDARDS

As covered before in this guide, other design standards exist. Listed below are some of the most commonly used standards in the UK. However, it should be noted that there are many other international standards.

Fire Offices' Committee rules

Although now obsolete, the Fire Offices' Committee (FOC) rules for automatic sprinkler installations were originally written in 1885. The last issue of these rules was published as the 29th edition in 1968 and this was withdrawn in 1990 when BS 5306-2[3] took over as the principal sprinkler standard. Some old FOC installations still exist and continue to offer a reasonable standard of protection.

LPC Rules incorporating BS 5306-2

BSI first published a code of practice for sprinkler systems in 1952 which was enlarged and superseded in 1979 by BS 5306-2:1979 *Fire extinguishing installations and equipment on premises. Sprinkler systems*.[15] BS 5306-2:1979 referred to the FOC rules. BSI undertook a complete revision of BS 5306-2: 1979 and embodied the full requirements of the FOC rules and other unpublished amendments, and published BS 5306-2: 1990 *Fire extinguishing installations and equipment in buildings – Specification for sprinkler systems*.[3] Updates, amendments and additional insurers' requirements, that were not included in this standard, were published as LPC technical bulletins.

The combination of the new British Standard (BS 5306-2: 1990[3]) and the LPC technical bulletins, formed the insurers' *LPC Rules* for automatic sprinkler installations.

LPC Rules for automatic sprinkler installations incorporating BS EN 12845

In October 2006, BS 5306-2: 1990[3] was superseded by BS EN 12845: 2003 *Fixed firefighting systems – Automatic sprinkler systems – Design, installation and maintenance* and was withdrawn.

EN 12845 was approved by CEN (the European Committee for Standardization) on 29 November 2002 and was introduced and became operational in the UK during 2003 (as BS EN 12845: 2003). The CEN members are the national standards bodies of Austria, Belgium, Czech Republic, Denmark, Finland, France, Germany, Greece, Hungary, Iceland, Ireland, Italy, Luxembourg, Malta, Netherlands, Norway, Portugal, Slovakia, Spain, Sweden, Switzerland and the United Kingdom.

The *LPC Rules* are divided into three main sections:
*   Part 1 contains the complete text of BS EN 12845: 2003. *Fixed firefighting systems – Automatic sprinkler systems – Design, installation and maintenance*.[2] † (see footnote on page 2).
*   Part 2 contains technical bulletins which explain and augment the requirements of BS EN 12845: 2003 and introduces innovative concepts such as the ESFR sprinkler.
*   Part 3 contains supplementary sprinkler information and guidance, including a guide to domestic and residential sprinklers.

The original *LPC Rules* incorporating BS 5306-2: 1979[15] ran in parallel with BS EN 12845[2] (published in 2003, then revised again in 2004) until 2006, when BS 5306-2 was withdrawn.

National Fire Protection Association of America

The National Fire Protection Association of America (NFPA) provides research, training, education, codes and standards for fire prevention. Established in 1896, NFPA's 300 codes and standards influence most building, process, service, design and installation in the USA and throughout the world.

The standard for the installation of sprinkler systems in the USA is NFPA 13.[16] There are other standards such as NFPA 13D[17] and NFPA 13R[18] for sprinkler systems in dwellings and residential occupancies up to four storeys in height.

NFPA standards separate fire pumps, water storage tanks, private fire service mains and most other major components into dedicated codes. Most codes and standards are expanded and updated with new publications at intervals of between one and four years.

Factory Mutual Global

Factory Mutual (FM) Global is one of the world's largest commercial and industrial property insurance and risk management organisations. It develops its own loss prevention practices and industry standards through independent research and testing. Many of the FM Global standards are based upon NFPA standards with additional requirements and recommendations.

FM Global names its standards 'datasheets'. The most commonly used datasheet is the 2-8N NFPA 13 *Standard for the installation of sprinkler systems*, 1996 edition.[19] However, like NFPA, FM Global also has separate datasheets for fire pumps, water storage tanks, suppression mode automatic sprinklers and many more.

In essence, most of the standards throughout the world provide a similar approach to sprinkler design and installation; however, there are some substantial differences between these different rules and standards which should be noted. Some differences between the two most common, the *LPC Rules* and NFPA standards, are noted in section 12.2.

12.2 COMPARISON BETWEEN LPC RULES AND NFPA STANDARDS

Types of sprinkler systems

There are six types of sprinkler systems identified in the *LPC Rules*:
* wet pipe
* dry pipe
* alternate
* pre-action
* deluge system
* subsidiary dry pipe or alternate extension (tail-end).

The types of sprinkler systems defined in the NFPA standards are:
* wet pipe
* dry pipe
* pre-action
* deluge
* combined dry pipe pre-action
* anti-freeze
* circulating closed-loop.

By definition, wet pipe, dry pipe, pre-action and deluge systems are similar in both the *LPC Rules* and NFPA standards. NFPA standards do not recognise alternate and subsidiary dry pipe or alternate extension (tail-end) installations and the *LPC Rules* do not recognise circulating closed-loop systems.

Spacing of sprinklers

There are many similarities in the requirements for the spacing of sprinklers between the *LPC Rules* and the NFPA standards (Table 1).

Protection of concealed spaces

The protection of concealed ceiling or floor spaces is a requirement of both the *LPC Rules* and the NFPA standards.
* *LPC Rules* require all spaces more than 0.8 m deep to be sprinkler protected. Spaces less than 0.8 m deep need to be protected only if they contain combustible materials or they are constructed with combustible materials.

* NFPA standards require all concealed spaces to be sprinkler protected unless they comply with a number of exceptions relating to construction and materials. If concealed ceiling spaces are not protected, the minimum design area for sprinkler operation is increased to 279 m².

Types of sprinkler head

There are many types, sizes and styles of sprinkler head. There is a large degree of similarity between sprinkler heads in the *LPC Rules* and NFPA standards.

Typical styles of sprinklers covered by the *LPC Rules* are:
* conventional (upright or pendant)
* pendant
* upright

Table 1: Maximum coverage and spacing for sprinklers (other than sidewall) for *LPC Rules* and NFPA standards

Maximum coverage and spacing for sprinklers (other than side wall)

Maximum area per sprinkler

	LPC (m²)	NFPA (m²)
Light hazard occupancy	21.0	20.9
Ordinary hazard occupancy	12.0	12.0
High hazard – process	9.0	9.3
High hazard – storage	9.0	9.3

Maximum coverage and spacing for side wall sprinklers

Maximum area per sprinkler

	LPC (m²)	NFPA (m²)
Light hazard occupancy	17.0	11.1/18.2
Ordinary hazard occupancy	9.0	7.4/9.3

Maximum coverage and spacing for sprinklers (other than side wall)

Maximum distance between sprinklers

	LPC (m)	NFPA (m)
Light hazard occupancy	4.6	4.6
Ordinary hazard occupancy	4.0	4.6
High hazard – process	3.7	3.7
High hazard – storage	3.7	3.7

Maximum distance between sprinklers

	LPC (m)	NFPA (m)
Light hazard occupancy	2.3	2.3
Ordinary hazard occupancy	2.0	2.3
High hazard occupancy	1.85	1.85

Maximum coverage and spacing for side wall sprinklers

Maximum distance between side wall sprinklers

	LPC (m)	NFPA (m)
Light hazard occupancy	4.6	4.3
Ordinary hazard occupancy	3.4	3.0

Maximum distance from side wall sprinkler to end of the wall

	LPC (m)	NFPA (m)
Light hazard occupancy	2.3	2.3
Ordinary hazard occupancy	1.8	1.5

- concealed
- flush
- recessed
- side wall
- extended coverage side wall
- ESFR
- dry pendant
- multi-jet control.

NFPA standards also cover extended coverage light hazard, extended coverage ordinary hazard, extended coverage high hazard and larger orifice types of extended coverage side wall sprinklers. The use of the conventional style sprinkler (old style in NFPA standards) is very limited in the NFPA 13 standard. There is no equivalent in NFPA standards for multi-jet controls.

Water supply types
Acceptable water supplies to NFPA standards are similar to those listed in the *LPC Rules*. There are two specific differences in the NFPA standards for water supplies:

- The *LPC Rules* require a minimum of two fire pumps for a superior water supply – one duty and one standby. NFPA standards will accept a single pump. Fire pumps constructed to NFPA standards, especially diesel engine driven fire pumps, are considered to be very reliable therefore NFPA standards only require a single pump.
- *LPC Rules* accept the arrangement of suction lift pumps whereby the pump is located above the water level and the pump suction line is primed with water from a tank and maintained by a foot valve. NFPA standards will not accept this arrangement and when pumps are located above the water level, they must be vertical turbine type.

Design densities and assumed areas of operation
The *LPC Rules* provide specific design densities and assumed areas of operation. The required design densities are similar for all types of sprinkler installations and are shown in Table 2.

The AMAO is the maximum area covered by a sprinkler installation that is assumed to be in simultaneous operation. The AMAO varies with the type of sprinkler installation. The AMAO for wet pipe or pre-action sprinkler installations is shown in Table 3.

For dry pipe or alternate sprinkler installations the AMAO is increased by 25%.

Design densities and assumed areas of operation vary for high hazard storage areas with sprinklers at roof level only or with in-rack sprinklers. They also vary for special hazard risks as defined in the *LPC Rules* technical bulletins.

NFPA standards provide for a selection from area/density curves where the design density will vary by the selection of the assumed area of operation (Table 4).

For dry pipe systems, the area of operation is increased by 30%.

Simultaneous operation of intermediate sprinklers in storage racks
The requirements relating to the design criteria for intermediate sprinkler protection varies considerably across the *LPC Rules* and standards.

In the *LPC Rules* the requirements for the simultaneous operation of in-rack sprinklers are:

- For hydraulic calculations, it is assumed that three sprinklers are in simultaneous operation at each level of in-rack sprinklers up to a maximum of three levels.
- The location of the three sprinklers and the three levels, when there are more than three levels of intermediate sprinklers, will be at the most hydraulically remote position.

Table 2: Specific design densities and assumed areas of operation in the *LPC Rules*

	mm/min.
Light hazard occupancy	2.25
Ordinary hazard group 1 occupancy	5.0
Ordinary hazard group 2 occupancy	5.0
Ordinary hazard group 3 occupancy	5.0
Ordinary hazard group 4 occupancy	5.0
High hazard process type/group 1	7.5
High hazard process type/group 2	10.0
High hazard process type/group 3	12.5

Table 3: The AMAO for wet pipe or pre-action sprinkler installations in the *LPC Rules*

	m²
Light hazard occupancy	84
Ordinary hazard group 1 occupancy	72
Ordinary hazard group 2 occupancy	144
Ordinary hazard group 3 occupancy	216
Ordinary hazard group 4 occupancy	360
High hazard process type/group 1	260
High hazard process type/group 2	260
High hazard process type/group 3	260

Table 4: Design densities and areas of operation in the NFPA standards

	Hazard design density selection (l/min/m²)	Area of operation selection (m²)
Light hazard	2.8 to 4.1	279 to 139
Ordinary hazard 1	4.1 to 6.1	372 to 139
Ordinary hazard 2	6.1 to 8.1	372 to 139
Extra hazard group 1	8.1 to 12.2	465 to 232
Extra hazard group 2	12.2 to 16.3	465 to 232

- Where rack aisles are 2.4 m in width or more, only one rack need be assumed in operation. Therefore, the maximum numbers of operating sprinklers to be calculated are three sprinklers at each of three levels, totalling nine sprinklers.
- Where rack aisles are less than 2.4 m in width and not less than 1.4 m, two racks will be assumed in operation. Therefore, the maximum numbers of operating sprinklers to be calculated are three sprinklers at each of three levels, in two racks, totalling 18 sprinklers.
- Where rack aisles are less than 1.2 m in width, three racks will be assumed in operation. Therefore, the maximum numbers of operating sprinklers to be calculated are three sprinklers at each of three levels, in three racks, totalling 27 sprinklers.

It is not necessary to calculate the simultaneous operation of more than three rows of sprinklers in the vertical plane or more than three rows of sprinklers in the horizontal plane, for each rack assumed in operation.

In NFPA 13, the requirements for the simultaneous operation of in-rack sprinklers are:

- Class I, II and III commodities – six sprinklers when only one level is installed in racks
- Class IV commodities – eight sprinklers when only one level is installed in racks
- Class I, II and III commodities – ten sprinklers (five on each top two levels) when more than one level is installed in racks
- Class IV commodities – 14 sprinklers (seven on each top two levels) when more than one level is installed in racks.

The minimum operating pressure for in-rack sprinklers is 1 bar. Some design criteria require a minimum operating pressure of 2 bar. K80 and K115 sprinklers are used in-rack and face sprinklers and barriers are required for some storage configurations.

Hydraulic calculations

The *LPC Rules* and NFPA standards use the same basic formulae for hydraulic calculations. The calculation of friction loss in pipework is that derived from the Hazen-Williams formula:

$$p = \frac{6.05 \times 10^5}{C^{1.85} \times d^{4.87}} \times L \times Q^{1.85}$$

Where:
p is the pressure loss in the pipe, in bar
Q is the flow through the pipe, in litres per minute
d is the mean internal diameter of the pipe, in millimetres
C is a constant for the type and condition of the pipe
L is the equivalent length of pipe and fittings, in metres.
Static pressure is calculated by:
$$p = 0.098\,h$$
Where:
p is the static pressure difference, in bar
h is the vertical distance between points, in metres

Water flow from a sprinkler is calculated from:
$$Q = K \times \sqrt{P}$$
Where:
Q is the flow in L/min
K is the constant (K factor)
P is the pressure in bar.

The C value for the most common types of pipes used in sprinkler systems are shown in Table 5.

In the *LPC Rules*, the water velocity is restricted to the following maxima: through any valve or flow monitoring device (6 m/s) and at any other point in the system (10 m/s). NFPA standards do not impose a velocity restriction through valves and monitoring devices.

The most significant difference between the *LPC Rules* and the NFPA standards is that the *LPC Rules* require hydraulic calculations for the hydraulically most unfavourable location and the hydraulically most favourable location. NFPA standards only require hydraulic calculations for the hydraulically most demanding design area. This is the same as the hydraulically most unfavourable location.

The *LPC Rules* use the point of intersection between the water supply curve and the system resistance curve derived from the hydraulic calculation of the hydraulically most demanding location, plus 0.5 bar, to determine the rating of the fire pump(s).

The intersection point between the water supply curve and the system resistance curve derived from hydraulically most favourable location is used to determine the maximum demand flow, and therefore the water storage requirements. NFPA standards base

Table 5: The value of C for the most common types of pipes used in sprinkler systems as set out in the *LPC Rules* and the NFPA standards	
LPC Rules	
Type of pipe	**Value of C**
Cast iron	100
Ductile iron	110
Mild steel	120
Galvanised steel	120
Spun cement	130
Cement lined cast iron	130
Stainless steel	140
Copper	140
Reinforced glass fibre	140
NFPA standards	
Type of pipe	**Value of C**
Unlined cast or ductile iron	100
Black steel (dry systems)	100
Black steel (wet systems)	120
Galvanised all	120
Plastic listed all	150
Cement lined cast iron or ductile iron	140
Stainless steel or copper tube	150

the required pump rating and water storage capacity on the hydraulically calculated demand for the hydraulically most demanding design area.

The jurisdictional authorities may require that an adequate safety margin is provided.

Combined systems

The *LPC Rules* do not permit the combination of sprinkler and hydrant systems. There is some limited potential to supply hose reel systems and non-industrial connections from town mains and on single storey buildings from stored and pumped supplies but, in general, only sprinklers may be fed from the supply.

NFPA standards permit the combining of sprinkler and hydrant supplies. The water supply is sized for the system with the greatest demand, satisfying the highest pressure and flow demands of the most demanding system. The combined system concept extends beyond the water supply and single mains can be used as the supply to combined system.

13 INSTALLERS AND INSTALLATION

One of the most important factors when considering the installation of sprinklers is the capability of the designers and installers. It is vital that those entrusted with this process are suitably trained and practised.

13.1 INSTALLERS AND SELECTION

The installation of certified sprinkler systems should be limited to suitable installation companies who have been assessed and approved under a recognised third party certification scheme. LPCB operates a scheme for approval of products as well as services listed in the LPCB Red Book which can be viewed free of charge at www. redbooklive.com.

There are four levels of LPS 1048[9] approval available. Companies without the necessary approval level are restricted in the type of work they are permitted to undertake. This is detailed in Table 6.

For companies undertaking only residential and domestic sprinkler systems a separate scheme is used: LPS 1301 *Requirements for the approval of sprinkler installers in the UK and Ireland for residential and domestic sprinkler systems.*[7]

13.2 RELIABLE INSTALLATION OF SPRINKLER SYSTEMS

The reliable installation of a sprinkler system is dependent on a number of main issues:

- The selection of a suitable installation company which is listed in the LPCB Red Book[13] or a scheme of equal status.
- Detailed design drawings, and where applicable hydraulic calculations, are prepared and issued for approval to the fire insurer and all other parties having either an interest or jurisdictional authority.
- Careful co-ordination with structural and mechanical, and electrical obstructions to ensure sprinkler heads are in the correct position to do their job, and pipe routes are appropriate.
- Careful inspection, testing and approval as fit for purpose for existing water supplies, where these are involved.
- Sprinkler components are suitable for their intended use and listed in the LPCB Red Book where applicable.

Table 6: LPS 1048[9] – levels of approval for sprinkler installers

Work type	Approval level			
	1	2	3	4
Systems, installations, extensions and alterations involving pre-calculated design principles	√	√	√	√
Base build contracts (pre-calculated design principles)		√	√	√
Systems, installations, extension and alterations involving fully hydraulically calculated design principles			√	√

- Materials used for pipes, pipe fittings and pipe supports are fit for purpose and comply with the applicable rules.
- Restrictions relating to the welding of pipes are strictly adhered to.
- Adequate care of materials onsite including the unloading and storing in a manner whereby they will not be damaged onsite. Pipes should be protected with caps to prevent the ingress of foreign matter, whether they are in storage onsite or erected.
- Sprinkler system should be correctly tested and commissioned.
- Sprinkler system is certified and an LPCB certificate of conformity or similar third party certification issued.

13.3 TESTING AND COMMISSIONING

Sprinkler systems are often subjected to progressive testing during installation. On installations equipped with flushing caps, an adequate volume of water should be used to flush any debris or foreign matter that may be in the pipes.

All pipework is required to be hydraulically tested to 15 bar or 1.5 times the working pressure, whichever is the greater, for a period of no less than two hours. Any failure during testing is to be corrected and the test repeated. It is often beneficial to undertake an initial pneumatic test in water sensitive areas.

The sprinkler system installer is committed to undertake a number of acceptance tests that authorities are invited to witness. These tests include, but are not limited to:

- Performance testing of all fire pumps, including specific testing of diesel engine driven pumps.
- Water flows and pressure at all installation valve locations that are remote from the water source testing facilities.
- Operation of air compressors and jockey pumps.
- Operation of all system alarm and fault signalling devices.
- Water storage tanks and pump priming tanks are to be commissioned full of water and infill rates must be proven where appropriate.
- All batteries are to be fully charged and diesel fuel tanks filled with diesel fuel.

13.4   CERTIFICATION

The jurisdictional authorities or the end user may require the sprinkler systems to be subject to an LPCB certificate of conformity. LPS 1048 certified sprinkler installers are required to issue certificates of conformity for all compliant sprinkler installations, extensions and alterations that conform to the *LPC Rules*. LPS 1301[7] certified sprinkler installers are required to issue certificates of conformity for all compliant residential or domestic sprinkler installations.

Only LPCB-approved contractors can issue LPCB certificates of conformity.

14 SERVICE AND MAINTENANCE

The service and maintenance of a sprinkler system is an essential activity to ensure the system will perform as initially designed, and to enhance the life of the system components.

14.1 OPERATION AND SERVICING

The user is required to carry out a programme of inspection and checks and to arrange for service and maintenance to be carried out on the sprinkler system.

The *LPC Rules* identify schedules of tests and inspections to be carried out weekly and at other specific frequencies, as well as precautions which should be taken while the work is being carried out. Weekly tests are normally those which may be carried out by the appointed staff of the user and include pressure and water level readings, valve position checks, testing of alarm functions and running and exercising pumps, where these are fitted.

Servicing and maintenance of the system and its components must be carried out quarterly, half-yearly and annually as required in the *LPC Rules*. Normally, for an LPCB certificated or similarly qualified contractor, would be appointed to carry out this work.

14.2 HAZARD VERIFICATION

It is essential to monitor any changes that may affect or change the hazard classification of a building protected by sprinklers. Sprinkler systems are designed to specific criteria based on the hazard classification of the building, and where applicable, the goods stored within.

Regular inspections should be carried out, and any changes noted and reported to the fire insurers. The inspection should include, but not be limited to:

- the classification of commodities
- change in packing method and materials
- change in storage method and heights
- movement or addition of internal walls, machinery and mechanical plant.

External storage of items, such as idle pallets that may cause an exposure hazard to the premises, should also be monitored.

14.3 RECORDS

Drawings, documents, hydraulic calculations, certificates, manufacturers' technical data, equipment manuals, operating and maintenance manuals, and relevant correspondence issued on completion of a sprinkler installation should be recorded. These drawings and documents should also be maintained and updated when any changes occur.

14.4 RECORD DRAWINGS AND DOCUMENTATION

It is essential to ensure that all relevant drawings and documents relating to the design of a sprinkler system are collated and passed to the building owner/occupier.

Record drawings in reproducible format, either as hard copy or electronic files, should be revised and issued in an 'as built' format. It is also important that clear copies of all hydraulic calculations are provided with reference drawings and hydraulic curves of the water supply and system resistance.

Other documents that should be provided on completion are: copies of manufacturers' technical datasheets and equipment manuals, copies of all certificates issued for compliance to standards and testing, and copies of all relevant correspondence with the fire insurer and other jurisdictional authorities.

14.5 OPERATING AND MAINTENANCE MANUALS

Operating and maintenance manuals should cover all aspects for the inspection, service, testing and maintenance of a sprinkler system. It is essential that data is provided to ensure that testing the fire pump performance, and water flow and pressure requirements for installation valve sets, is matched to the performance achieved at commissioning. Recommendations for service and maintenance issued by manufacturers of fire pumps, air compressors etc., should be undertaken.

The following items are the minimum expected documentation required in a sprinkler system operation and maintenance manual:

- project name and contract details, including the sprinkler contractor

- full description of system
- confirmation of design criteria, and any special bespoke requirements by third parties
- health and safety considerations
- procedures in the event of a fire or emergency
- detailed description of the system operating, testing and maintenance procedures and requirements
- system fault finding
- catalogues and technical data for all equipment used, especially the water supplies

- name, address and contact numbers of all suppliers for spare parts
- test certificates
- completion certificates
- as-fitted design drawings and hydraulic calculations
- block plan
- emergency contact numbers
- water supplies test data.

15 CHANGES OF USE

The change of use of an existing building is a common occurrence, eg the ownership/tenancy of a building may change.

Refurbishment and redevelopment of buildings can completely change the fire risk and classification of the hazard and thereby undermine the capabilities of the system. When such changes of use occur in sprinklered buildings, a reassessment of criteria associated with risk and classification should be undertaken to determine the suitability of the sprinkler system.

15.1   CHANGE IN RISK

Building alterations and additions could cause a change in risk. For example, if a large storeroom is added to an office building the risk is likely to increase.

Changes in the raw materials for processing and manufacturing can often change the risk as they may represent a higher combustibility.

15.2   CHANGE IN HAZARD CLASSIFICATION

A change in the manufacturing technique, raw materials, manufactured goods, type of packaging and method and height of storage, can all cause a change in hazard classification.

Typical examples are where a manufacturing technique is changed eg where different machines are used and changes in shop layouts – the introduction of painting could also change the area classification. In a commercial building the occupancy of areas may change and cause classification to move from an ordinary hazard to a high hazard.

A change in the composition of raw materials used in a manufacturing process may again change the hazard classification of the risk. For example, in a printing works the change of weight of paper would probably limit the allowable height of its storage.

The method of packaging goods on pallets, either individually or collectively, can change the hazard classification as well as change the storage height or method of storage. The introduction of shelved storage racks is often the cause of a re-evaluation of the hazard classification.

15.3   NOTIFICATION OF CHANGES

All changes to the building, hazard classification and sprinkler system must be notified to the fire insurers, fire officer and all other jurisdictional parties. If a sprinkler installation or zone of a sprinkler installation is taken out of service for repair or extension, the fire insurer and fire officer should also be notified.

16 APPROVALS

16.1   LISTING OF SPRINKLER PRODUCTS

To ensure that a sprinkler system is fully compliant with the applicable design standards (and in the case where the design standard is LPC and an LPCB certificate of conformity is to be issued), all appropriate sprinkler products must be listed and approved by a third party certification body such as LPCB.

16.2   LISTED PRODUCTS

LPCB publishes a book of approved fire and security products and services, which is updated at regular intervals. This publication is for use in conjunction with the *LPC Rules* and it covers details, equipment and services (which have been assessed for quality and performance) approved by LPCB.

The latest information and approval status are available from www.redbooklive.com.

16.3   LISTED INSTALLERS

The LPCB Red Book[13] provides the listing of companies assessed and certified by LPCB to be suitable installers of sprinkler systems. The listing of installers identifies the extent of their approved services.

The main categories are:
* LPS 1048 *Requirements for the approval of sprinkler system contractors in the UK and Eire.*[9]
* LPS 1301 *Requirements for the approval of sprinkler installers in the UK and Ireland for residential and domestic sprinkler systems.*[7]

See www.redbooklive.com for more details.

16.4   VARIANCES IN STANDARDS, LISTINGS AND APPROVALS

Sprinkler products, equipment and devices are subject to testing and listing by internationally accepted authorities. In the UK, testing, approving and listing are undertaken by LPCB. There are also many other approving authorities in Europe, USA and throughout the world.

In the USA, there are two significant approving authorities: Underwriters Laboratories Inc. (www.ul.com) and FM Approvals (part of FM Global www.fmglobal.com). There are subtle differences between Underwriters Laboratories Inc. and FM Approvals listings. For example, an ESFR sprinkler may be listed for differing design criteria, or the FM Approvals design standard requirements are more stringent than those listed by Underwriters Laboratories Inc.

There are many products that are listed by multiple approval bodies, such as LPCB, Underwriters Laboratories Inc. and FM Approvals, and likewise products that are only listed by one agency.

In the design of a sprinkler system, it is essential that all sprinkler products are listed in accordance with the applicable design standard, and by the jurisdictional authority for the design criteria.

REFERENCES

[1] Fire Protection Association, 2007. LPC Rules for automatic sprinkler installations incorporating BS EN 12845. Moreton-in-Marsh.

[2] BSI, 2003. BS EN 12845. Fixed firefighting systems. Automatic sprinkler systems. Design, installation and maintenance.

[3] BSI, 1990. BS 5306-2. Fire extinguishing installations and equipment on buildings – Specification for sprinkler systems.

[4] Communities and Local Government, 2006. The Building Regulations 2000. Approved Document B. Approved Document B – Volume 2 – Buildings other than dwellinghouses (2006 edition).

[5] Department of Health. Firecode: Fire safety in the NHS.

[6] BSI, 2005. BS 9251. Sprinkler systems for residential and domestic occupancies. Code of practice.

[7] BRE Global, 2007. LPS 1301. Requirements for the approval of sprinkler installers in the UK and Ireland for residential and domestic sprinkler systems. www.redbooklive.com/pdf/LPS1301-1.pdf.

[8] Department for Children, Schools and Families, 2008. SSLD 8 Sprinklers in schools, Standard specifications, layouts and dimensions. www.teachernet.gov.uk.

[9] BRE Global, 2003. LPS 1048 Requirements for the approval of sprinkler system contractors in the UK and Eire. www.redbooklive.com/pdf/LPS1048-1_Issue_4.pdf.

[10] BSI, 2008. BS 5839-1:2002+A2. Fire detection and fire alarm systems for buildings. Code of practice for system design, installation, commissioning and maintenance.

[11] BSI, 2005. BS EN 54-14. Fire detection and alarm systems. Guidelines for planning, design, installation, commissioning, use and maintenance.

[12] BSI, 2002. BS 6266. Code of practice for fire protection for electronic equipment installations.

[13] BRE Global. LPCB Red Book. List of approved fire and security products and services. www.redbooklive.com.

[14] Fire Protection Association, 1999. Technical briefing note: Legionella and firefighting systems. Moreton-in-Marsh.

[15] BSI, 1979. BS 5306-2. Code of practice for fire extinguishing installations and equipment on premises (sprinkler systems).

[16] National Fire Protection Association of America, 2007. NFPA 13: Standard for the installation of sprinkler systems. Massachusetts.

[17] National Fire Protection Association of America, 2007. NFPA 13D: Standard for the installation of sprinkler systems in one- and two-family dwellings and manufactured homes. Massachusetts.

[18] National Fire Protection Association of America, 2007. NFPA 13R: Standard for the installation of sprinkler systems in residential occupancies up to and including four stories in height. Massachusetts.

[19] Factory Mutual Global, 1996. Property loss prevention data sheets. Datasheet 2-8N: Standard for the installation of sprinkler systems. Johnston.